Y0-BOA-247

SPACE LIBRARY

SPACE LABORATORIES

GREGORY VOGT

FRANKLIN WATTS

NEW YORK LONDON TORONTO SYDNEY

Each new generation of humankind has had the challenge of a frontier. The frontier for today's children is outer space; it beckons with unlimited experiences. It is the frontier of my children, and I dedicate this book to them.

Kirsten, Allison and Catherine Vogt

First published in the USA
by Franklin Watts Inc.
387 Park Ave. South
New York, N.Y. 10016

First published in Great Britain in 1989 by
Franklin Watts
96 Leonard Street
London EC2A 4RH

First published in Australia
by Franklin Watts
Australia
14 Mars Road
Lane Cove, NSW 2066

US ISBN: 0 531 10404 4
UK ISBN: 0 86313 598 6
Library of Congress
Catalog Card No: 87-50168

Printed in the United States of America

Designed by Michael Cooper

Cover photo courtesy of NASA.

All photographs courtesy of NASA except U.S. Naval Observatory: p. 6 (bottom); Novosti/ Science Photo Library: p. 7 (top right); Sovfoto, pp.12-13; European Space Agency: pp. 14-17; Mark Ponto, St. Mary's Hospital, Milwaukee, Wisconsin: p. 23 (top left).

CONTENTS

LABORATORIES IN SPACE

The business of scientists is to ask questions: How many stars are there in space? Why do plants' leaves grow up while their roots grow down? What happens when a force is exerted on an object? These and millions of other questions have occupied scientists for hundreds of years. Until recently, scientists had one thing in common: They all studied the Universe from the same laboratory—the surface of the Earth.

The Earth was an adequate laboratory for most studies, but as questions were answered, new questions were asked: How many stars could be seen with a telescope placed above the atmosphere of the Earth? Would plants still grow the way they do if gravity were not affecting them? How would an object move if a force were exerted on it in outer space? But there was a problem: Answering these questions required a new laboratory, one not on the surface of the Earth. The new laboratory would have to be in outer space.

The problem was how to get it there. Until that problem was solved, scientists could only theorize about the results of their experiments in space. Astronomers were sure they would be able to see fainter and more distant objects through the vacuum of space. The Earth's atmosphere is not nearly as clear as they would like it to be. Clouds and haze obscure their view, and they could only work at night. Even on very clear nights heat currents in the atmosphere produce a shimmering effect on the light entering their biggest telescopes.

Physicists wanted to test their ideas about force and motion. But friction produced by gravity interfered with many of their studies. Theoretically, a satellite could be made to orbit the Earth if it were moved fast enough parallel to the ground. The satellite would actually be in a state of freefall, and the objects inside would be weightless.

One of the captured German V2 missiles lifts off, carrying a Wac/Corporal rocket on top in a U.S. test of a two-stage vehicle in the late 1940s.

Taken through an Earth-based telescope, this photo of galaxy NGC 7331 only hints at its structure. A telescope placed above the filtering veil of Earth's atmosphere would reveal much more detail.

Weightlessness offers a new dimension for scientific experimentation. Astronaut Allen Bean enjoys acrobatics in the Skylab space station. A simple push against a wall and he can tumble for hours.

Bean seedlings exhibit unusual root growth in this 1985 Space Shuttle plant growth experiment.

Biologists wanted to place living things in orbit in weightless laboratories to find out, among other things, if leaves grow up and roots grow down in an environment where there really is no up or down. Perhaps the reverse would be true. Chemists wanted to know if liquids of different densities would stay mixed in space. Doctors wanted to know if the lack of up or down in space would confuse people so much that they might think they were going crazy. Answering these and many other questions required traveling into space.

The solution to the problem of how to conduct scientific investigations in outer space was found, surprisingly, in the destructive forces of World War II. Powerful V2 rockets had been built by Germany to carry explosive warheads to its enemy Great Britain. After the war, captured V2 rockets and German rocket scientists like Wernher von Braun were brought to the United States to help develop the growing U.S. space program. The launching of V2s provided U.S. scientists with valuable information that they would soon need. Other German rocket technology and rocketry experts went to the Soviet Union.

THE FIRST SPACE LABS

On October 4, 1957, the Soviets took the first step in what would soon be known around the world as "the space race." A thundering rocket carried Sputnik 1, the world's first artificial satellite, into space, signaling to scientists everywhere the start of a new era. The new laboratory of outer space was open for business and the whole world eagerly awaited the results.

The early satellites were primitive affairs. The first generation were little more than containers into which a transmitter was placed, along with a power supply and a few scientific instruments. The instruments themselves were also very basic: a Geiger counter for measuring radiation, a microphone to listen for the *pings* of tiny meteoroids bouncing off the shell, light sensors, and thermometers.

Although simple in design, these satellites told scientists much about the environment of outer space.

One of the important discoveries of the early days of satellite launching was the bands of radiation circling the Earth. Their presence was detected by a Geiger counter placed on board the U.S. Explorer 1 satellite by scientist James Van Allen. By measuring the path of the satellite and the strength of the radiation it encountered, Van Allen determined that two large doughnut-shaped regions of space surrounding the Earth are composed of electrically charged particles ejected from the Sun. These particles are trapped there by the Earth's magnetic field. In his honor, these regions were named the Van Allen Radiation Belts.

Not long after the first satellite launch, the first living creature from Earth was launched into orbit, on November 3, 1957. The Soviet Sputnik 2 satellite was designed by Russian scientists as a laboratory in space. The large satellite weighed 508 kg (1,120 lb) and consisted of two spheres and a cylinder placed inside a large tube.

(Left) Weather scientists were elated by space views of cloud tops transmitted by early weather satellites like TIROS II, launched on November 23, 1960. (Below) The Van Allen Radiation Belts were an important discovery about the Earth's near-space environment made by the first U.S. satellite, Explorer 1. Electrically charged particles from the Sun are trapped by the Earth's magnetic field, forming two ring-shaped regions circling the Earth.

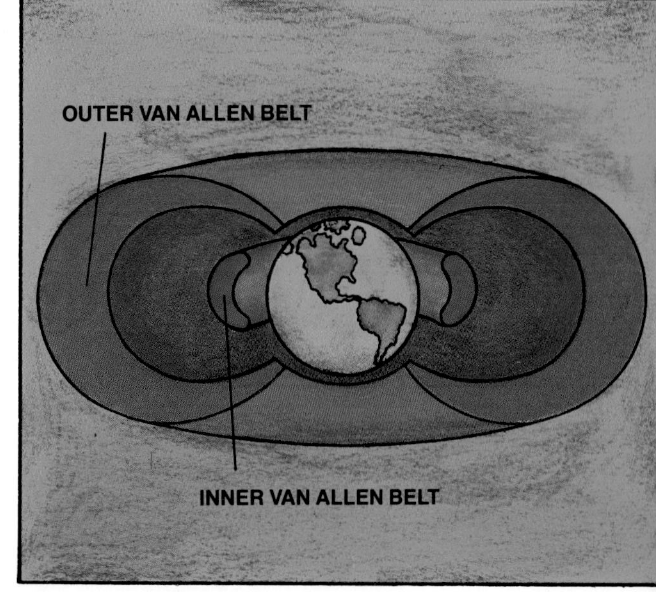

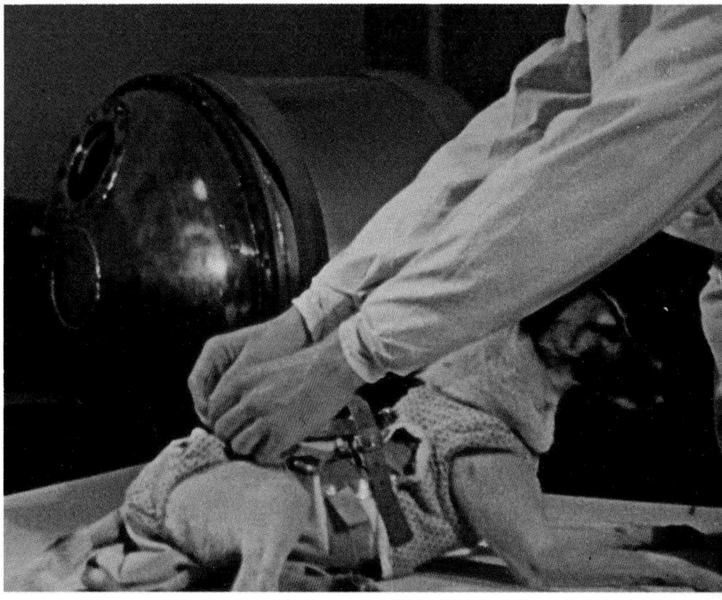

(Left) One of the five chimpanzees used for Mercury-Atlas rocket flights poses in a netted flight suit near the flight couch on which he will ride during the flight. Chimpanzees were used to prove that space flight was safe for humans. (Above) Laika, the first animal to orbit the Earth, readied for lift-off. Laika traveled on board the Soviet Union's second satellite.

Inside Sputnik 2 was a female dog named Kudryavka ("Little Curley"), but most people called the dog Laika, after the name of her breed. Laika survived in space for several days, proving that living things are not harmed by weightlessness. Unfortunately, food and air for only a few days were provided for Laika. There was no reentry capsule for the dog's return, so she was killed to prevent her from suffering. Laika's ability to survive inside a satellite paved the way for the first human in space three and a half years later.

Following Laika's flight, there were more Soviet scientific experiments involving dogs. By this time, the Russians had learned how to bring the animals back safely and there was little danger that the animals would perish. Meanwhile, the National Aeronautics and Space Administration (NASA) in the United States was preparing its first animal experiments in space. NASA chose monkeys and chimpanzees for their flights because they more closely resembled humans.

One especially interesting satellite experiment was launched on November 29, 1961; it included a chimp called Enos. Enos's satellite was a Mercury spacecraft of the same kind in which John Glenn would later travel into space. Like an astronaut, Enos rode in the capsule seat. He even had jobs to do, like pulling levers on cue from the ground. If he performed his job correctly, he was given banana pellets and water. If he made a mistake, he was given a mild electrical shock to his feet.

Unfortunately, Enos's flight was plagued with problems. The spacecraft began to tumble, and the mission had to be ended early. The cooling system broke down, and it became very hot inside the craft. To add further insult, a circuitry malfunction sometimes gave him shocks even when he pulled the correct lever. After splashdown, Enos jumped up and down on the deck of the recovery ship, perhaps in anger or in relief at being released from that awful space capsule. Nevertheless, Enos added to the knowledge of space flight by demonstrating that living creatures could survive the hostile atmosphere of outer space, at the same time doing constructive jobs.

LUNAR LABORATORY

Much of the world watched and thrilled at the Apollo astronauts as they walked on the Moon, but most seemed to forget that the astronauts were there for a purpose. Our closest neighbor in space, the Moon offers a unique environment. Unlike the Earth, the Moon is airless. It is boiling hot in the sun and terribly cold in the shade. Lacking the shield of an atmosphere, it is bombarded by cosmic radiation and X rays. Without air and the accompanying water, there is little erosion to obliterate the impact craters that are left by meteors and asteroids that strike the Moon's surface, even ones that arrived millions or billions of years ago. The Moon offered the Apollo astronauts an excellent laboratory for studying the history of our Solar System.

To prepare for manned missions to the Moon, NASA sent three different kinds of robot spacecraft to the Moon. Ranger spacecraft were designed to crash into the Moon; just before doing so, they sent back close-up pictures. Surveyor spacecraft landed on the Moon to study its surface structure. And Lunar Orbiters mapped the Moon's surface photographically, including the far side, so that landing sites could be selected.

The spacecraft completed their survey work, and on July 20, 1969, the first of six Apollo Lunar Modules (or LMs, pronounced *lemz*) touched down on the Moon. After landing, the primary task of each mission was to collect samples of lunar soil and rock. The astronauts were especially good at this, and 383 kg (843 lb) of samples were brought back to scientists on Earth for study. Some of the rocks were found to be more than four billion years old.

Another important part of each Apollo mission was to set up scientific experiments on the Moon's surface. Some experiments were very simple, such as placing reflectors on the surface to bounce back laser beams sent from Earth. By timing how long it takes for the beam to return, scientists could calculate the distance between the Moon and the Earth to within a few centimeters. Such accuracy had previously been impossible.

Harrison Schmitt standing next to a large boulder during a Moon walk on the Apollo 17 mission. His lunar rover is seen in the distance to the right.

Scientists have sliced Moon rocks so thin that light can pass through them. In this view through a microscope of a sample of basalt rock brought back by the Apollo 12 crew, crystals of minerals become clearly visible.

Many other experiments were much more complicated. Starting with the second Apollo mission, each crew set up an automatic science laboratory called the Apollo Lunar Scientific Experiments Package (ALSEP). At its head was an atomic power supply; this sent electricity through tentacle-like cables to widely spaced experimental stations. Experiments at these stations measured the Moon's magnetic field, its internal heat, and moonquake activity. Other experiments studied the interaction between the particles ejected by the Sun and the Moon's surface.

The Apollo expeditions to the Moon provided scientists with a treasury of data and samples that are still being analyzed. In one especially interesting and bizarre set of experiments, discarded portions of a lander and a launch vehicle from Saturn V were targeted to crash on the Moon's surface. Detectors could then listen to the vibrations and learn about the Moon's interior. Scientists were greatly surprised to find a much deeper layer of broken rock on the Moon's surface than anyone had previously assumed.

(Left) Apollo 11 astronaut Buzz Aldrin sets up a solar wind collector in an attempt to capture electrically charged particles ejected from the Sun.
(Below) After the Apollo 14 crew returned home, the Passive Seismic Experiment Package (Moonquake detector) continued collecting data and transmitting it back to Earth by radio.

SKYLAB, THE FIRST U.S. LAB

The Skylab space station was unlike any other U.S. spacecraft that had ever flown. It was designed as a permanent space laboratory, housing a crew of astronauts. Inside it divided into three "rooms," with a total of 390 cu m (13,000 cu ft) of space. The astronaut crews who lived and worked there spent most of their time conducting scientific experiments. Three crews of three astronauts spent a total of 169 days in space, arriving and returning in Apollo capsules.

Skylab provided facilities for a wide variety of investigations into human space flight, physics, chemistry, astronomy, biology, geology, meteorology and oceanography. Many medical experiments were conducted on the crew members themselves to learn about the long-term effects of space flight. In one experiment, crew members placed themselves feetfirst into a cylinder. Air was pumped out to reduce the pressure around the waist and legs. Sensors stuck on to various parts of the body determined what kinds of changes were taking place in the heart and the blood vessels. Every third day or so, each crew member became a test subject.

In another experiment astronauts used a special exercise bicycle called an ergometer. While pedaling, the astronaut breathed carefully measured gases so that scientists could learn the body's use of energy in space. The ergometer was similar to exercise bikes on the ground except that there was no seat. In the weightless condition of orbit, a seat was unnecessary.

(Below) The Skylab space station became a science laboratory and home to three crews of three astronauts from May 1973 to early February 1974.
(Left) The last crew of Skylab astronauts took this picture of the space station from their Apollo space capsule during a "fly around" inspection.

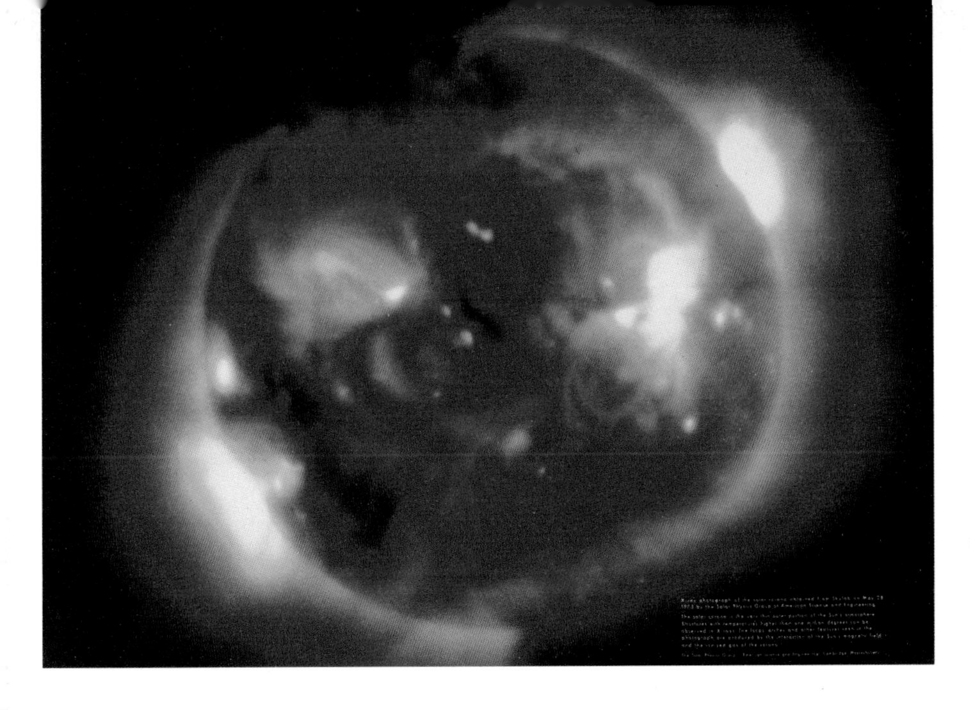

One of the 180,000 photographs of the Sun made by the Apollo Telescope Mount shows the Sun in X-ray light.

Does an astronaut gain or lose weight in space? One of the problems in answering this question is that bathroom scales don't work in weightless conditions. The solution was to use an oscillating chair to determine mass. A crew member would sit in the chair and release a spring that vibrated the chair back and forth. It takes energy to stop a crew member from moving in one direction and make him or her move in another, even though weightless. The more body mass present, the more energy it takes. In this manner, it was possible to measure the astronaut's mass.

Another group of experiments used telescopes to study the Sun both in ordinary light and in X-ray light. In some studies, the Sun was blocked, so that its atmosphere became visible. During the three Skylab missions, more than 180,000 pictures of the Sun were taken. These helped astronomers discover many new things about the Sun that could not have been observed through their telescopes on the ground.

Special telescopes and cameras provided exceptional views of the Earth from Skylab's 450-km- (278-mi-) high orbit. Mountains, plains and deserts were plainly visible. Surface colors gave subtle clues of mineral and oil deposits.

Still other experiments studied how crystals grow in the weightless condition of space and what liquids do when allowed to float freely. Globules of water floated around, looking like water balloons without the balloons. The astronauts had to resist almost overwhelming temptations to have water fights because of the danger of causing shorts in electrical equipment. One slight touch to the globules caused them to shimmer and vibrate.

When it was launched in 1973, Skylab was expected to last for many years in space. However, the Sun underwent a period of intense activity that released high levels of energy into the Solar System. The upper atmosphere of Earth warmed up and expanded so that traces of the atmosphere reached Skylab's orbit. Though slight, these collisions with the air caused it to slow down. In July 1979, the huge space laboratory reentered the Earth's atmosphere and disintegrated.

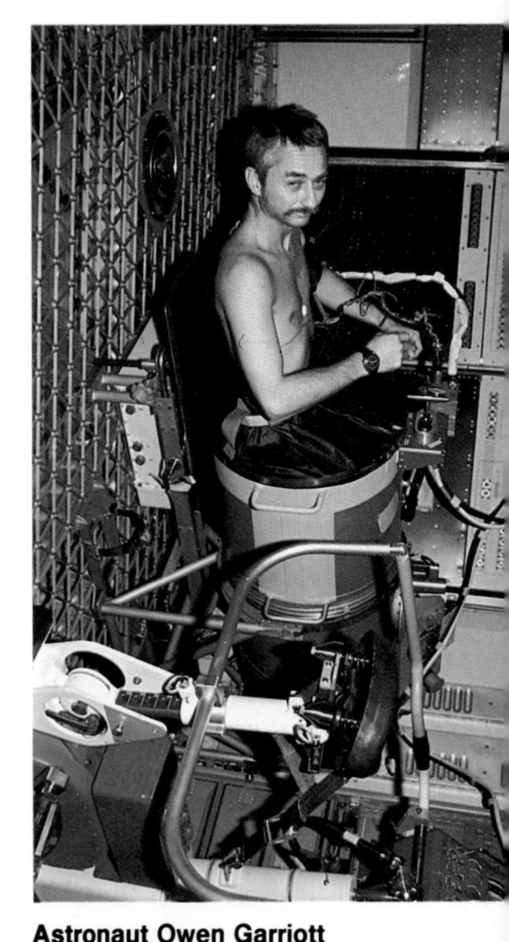

Astronaut Owen Garriott underwent tests on Skylab 3. The Lower Body Negative Pressure device in which he has placed himself helped evaluate his heart and his circulatory system's adaptation to space flight.

PERMANENTLY MANNED STATION

The loss of Skylab caused the U.S. space station program to grind to a halt. Now, ten years later, there still hasn't been a replacement for Skylab. NASA is working on a new space station but it probably won't be in operation before the mid-1990s, or later. The Soviet Union, on the other hand, has concentrated on space station work. Soviet scientists began with a station called Salyut 1, launched on April 19, 1971. They have orbited many more stations and presently Salyut 7 is in orbit along with a new generation station known as Mir, the Russian word for Peace.

Though none of the operating Soviet stations was or is as large as Skylab (Mir could just about fit in the Space Shuttle payload bay), they have been highly versatile space laboratories given over to particular missions. Some focused on civilian research, such as biology and chemistry and studies of the Earth from above. Others have concentrated on military research, perhaps to monitor strategic areas on Earth.

Mir, the most recently launched Soviet space station, is 17 m (56 ft) long and 4 m (13 ft) in diameter. It has two very large solar panels that look like giant seagull wings. Mir features six docking ports for linking with Soyuz manned space capsules and Progress unmanned supply capsules. Some of the ports will be used for attaching additional cylindrical modules that will be sent up in the future. Each of them is expected to be dedicated to a unique purpose such as specialized laboratories for biology, materials processing, astronomy and experiments in making new materials in space.

Inside, the main body of the station is relatively open and much of the space is used for crew quarters and as a control center.

Salyut 7 in orbit in 1986.

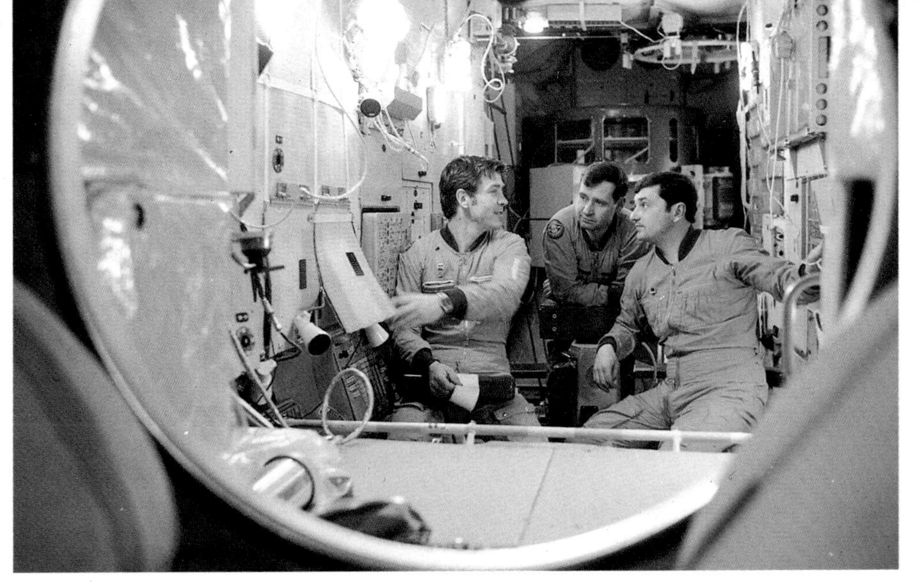

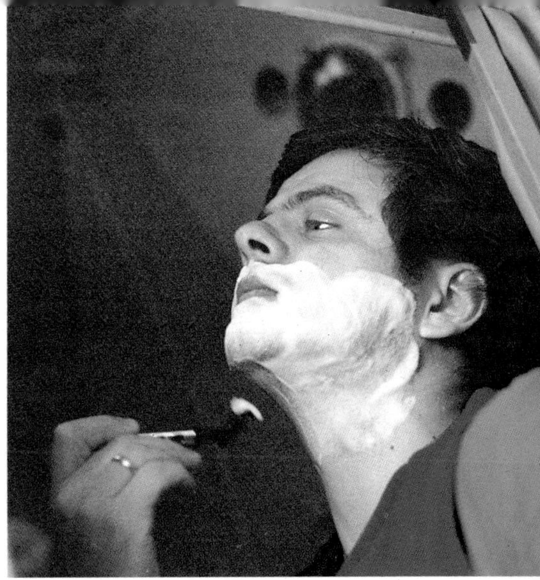

(Left) Crew members of the Soyuz T-8 at the simulator of the Salyut orbital station at the Gagarin Training Center. (Above) Cosmonaut V. Solovyov shaves aboard Mir.

Mir is designed for long stays in space. The Soviets hope to keep it resupplied with crew, oxygen, food and propellant so that it will be permanently occupied. Already two Soviet cosmonauts, Vladimir Titov and Moussa Mamarov, have remained on board Mir for a full year before returning to Earth. Because they stay for a very long time in space, the cosmonauts that occupy Mir have become human guinea pigs. They are given periodic medical tests to find out what changes are taking place in their bodies and to make sure they remain healthy. So far, most changes detected have been temporary ones, such as puffiness in the upper part of their bodies due to blood pooling. On Earth, blood in the human body is more or less uniformly distributed. In space weightlessness the blood tends to concentrate in the upper torso and head region. Another temporary change is the lengthening of the spine due to a lack of gravitational pressure. Astronauts experiencing this have been known to grow a few centimeters only to lose them once back on Earth.

Information gained about human reactions to long stays in space will come in handy. The Soviets and the U.S. are considering sending cosmonauts on interplanetary flights to Mars that might last as long as two years before returning to Earth.

On the ground, Mir is tested at the Baikonur Testing Station.

SPACELAB

One great advantage that the U.S. Space Shuttle has over most other space vehicles is its ability to carry payloads in a large cylindrical bay, 18 by 4.6 m (60 by 15 ft) in size. Another advantage is that it can return the payload back to Earth. Most other launch vehicles remain in space or make fiery reentries; if any part of them did return, it was usually a small space capsule with a few astronauts and a small payload inside. The Shuttle's big launch-and-return payload capacity encouraged many space groups to consider how to build substantial payloads. One group decided to build a space laboratory that scientists could actually work in. The group was the European Space Agency (ESA), and they named their laboratory Spacelab.

Spacelab was a complicated project. It was to ride in the Space Shuttle's payload bay; that bay's doors are designed to be open during orbital operations, so the laboratory had to be able to withstand the widely ranging temperatures and airlessness of space. The lab would also need a connecting passageway from the crew quarters, and a place to put instruments and experiments outside when needed. It soon became apparent to the designers that Spacelab would be not one large piece of equipment but a group of components that could be assembled in different ways to form a laboratory. During ten years of design, construction and testing, Spacelab took shape.

(Left) Kennedy Space Center technicians move Spacelab and a pallet during its processing for flight on board the Space Shuttle.

Technicians examine the subfloor assembly of Spacelab before the laboratory interior is installed.

A train of pallets and the laboratory module, covered with thermal insulating blankets, are arranged for shipment to the Kennedy Space Center.

The actual laboratory in which scientists work is divided into two cylindrical aluminum segments, each 2.7 m (9 ft) long by 4 m (13 ft) in diameter. The first segment is called the Core Module; on some space missions it flies alone. When that happens, it is also called the Short Module. The second segment, called the Experiment Module, can be joined with the core when more experiment space is needed. When the two are used together, the configuration is called the Long Module. The inside of the Long Module is like the passenger cabin of a large jet, but instead of seats and windows, the walls are lined with racks of instruments. Electric power to run all the instruments is provided by fuel cells carried by the orbiter.

Connecting the Long Module with the orbiter cabin is a 1-m-(3-ft-) diameter tunnel that has a Z-shaped bend at one end. This is necessary because the Shuttle orbiter returns to Earth as a glider.

The last major component of Spacelab is the pallet. This is a large U-shaped experiment rack that cradles scientific instruments that have to be exposed directly to outer space. Pallets have a mass of 1,200 kg (2,650 lb) and are 4 m (13 ft) wide and 3 m (10 ft) long. The experiments either run automatically or are controlled by the scientists inside the laboratory.

GETTING SPACELAB READY

Like the Space Shuttle, Spacelab is highly versatile. It has to be, because scientists are very demanding. The moment an experiment is over, they want to fly it again or change it or do something different. They're never satisfied, and Spacelab has to be ready to adapt to their needs. It does so through its flexible modular design.

Spacelab can fly in several different configurations to meet the special needs of each mission. It can fly with only its Short Module and two or three pallets. It can fly with the Long Module alone, or with one or two pallets, and up to five pallets can fly by themselves. These variations offer scientists countless possibilities.

There is also versatility within the Short and Long Modules. These modules are sometimes also called the Habitation Modules because scientists actually live and work inside them when they are in space. Inside, from one end to the other, the walls and ceiling form an arch. Actually, floor, ceiling and walls are words that apply only when Spacelab is on Earth being prepared for a mission. In space, there is no sense of gravity and no up or down. That is why handrails are found everywhere inside the modules. To make things easier for the scientists, all read-out dials and controls are marked so that they can be read from one direction, as if up and down really existed.

Spacelab 1 is installed in Space Shuttle orbiter *Columbia*'s payload bay. The tunnel that connects the lab to the orbiter mid deck is easily seen.

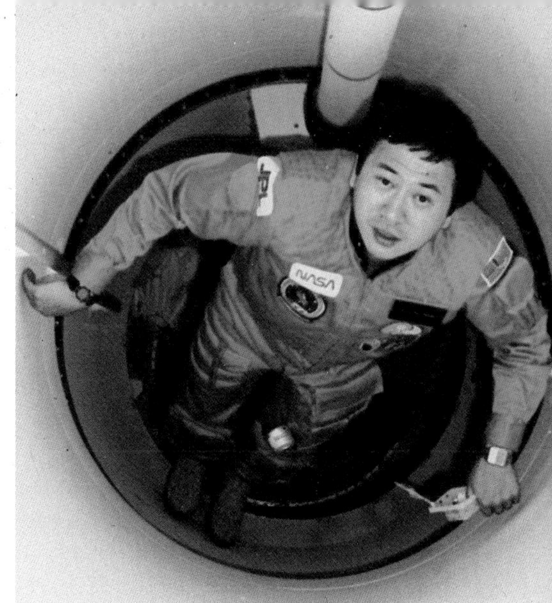

(Left) Inside view of Spacelab. (Above) Mission specialist Taylor Wang floats through the tunnel connecting the orbiter mid deck with Spacelab during the 51B mission.

Small experiments and measuring instruments slip into slots and are connected to the electricity and computer control systems. The powerful computer system can store 132 million bits of data—the equivalent of all the words in twenty-seven thick books. Everything is designed for convenience and compactness.

Overhead in the Core Module are two windows through which cameras can photograph oceans, clouds and land when the Shuttle's payload bay is aimed toward Earth. The Experiment Module has an airlock in its ceiling.

Experiments are also mounted on one or more pallets in back of the Habitation Modules. When an instrument such as a telescope must be precisely aimed at a target in space, an instrument pointing system (IPS) is used. The IPS motor drive is so precise that it could keep a gun, mounted on a moving automobile, sighted on a small marble from a distance of 3.6 km (2.4 mi) away. In space, it must compensate for the Shuttle's orbital movement of thousands of kilometers per hour and for the jiggling produced by the crew bouncing around inside.

When all the experiments and supplies are mounted inside the modules or on the pallets and have been tested and retested and tested again, the modules and pallets are hoisted into the orbiter's payload bay and mounted firmly to it.

A cutaway drawing showing the Spacelab laboratory and two pallets on board the Space Shuttle.

SPACELAB AT WORK

Ten years of design and construction paid off when Spacelab 1 flew on board the Space Shuttle *Challenger* on November 28, 1983. In addition to the usual Shuttle mission commander and pilot, four payload specialists were on board specifically to operate the seventy-two experiments planned for the mission. Included in the crew were NASA astronauts Owen Garriott and Robert Parker and scientists Byron Lichtenberg and Ulf Merbold. Merbold was on board on behalf of the European Space Agency; he came from the Federal Republic of Germany.

For ten and a half days, the Spacelab crew busied themselves with many different scientific experiments. Some of the experiments involved using materials in space. Crystals of mercury iodide were grown inside mini–vacuum chambers; a protein crystal of lysocyme was grown that was a thousand times larger than those produced on Earth. It was large enough to enable scientists on Earth to expose it to X rays to begin to understand its atomic structure.

While Spacelab orbited the Earth from an altitude of 240 km (146 mi), a metric camera on board was able to take two thousand pictures of the Earth's land surface. The pictures covered a total of 11 million sq km (4.3 million sq mi) and were so detailed that researchers could pick out medium-size buildings on the ground. The pictures were given out to 141 researchers worldwide in order to process them for maps.

Further scientific experiments measured the amount of methane in the Earth's atmosphere 70 km (43 mi) up and the amount of deuterium, a special form of hydrogen, 110 km (67 mi) up. Astronomers looked at regions of the sky in X-ray and ultraviolet light. Other experiments involved the crew members themselves in very practical studies to find out why many astronauts suffer from space sickness.

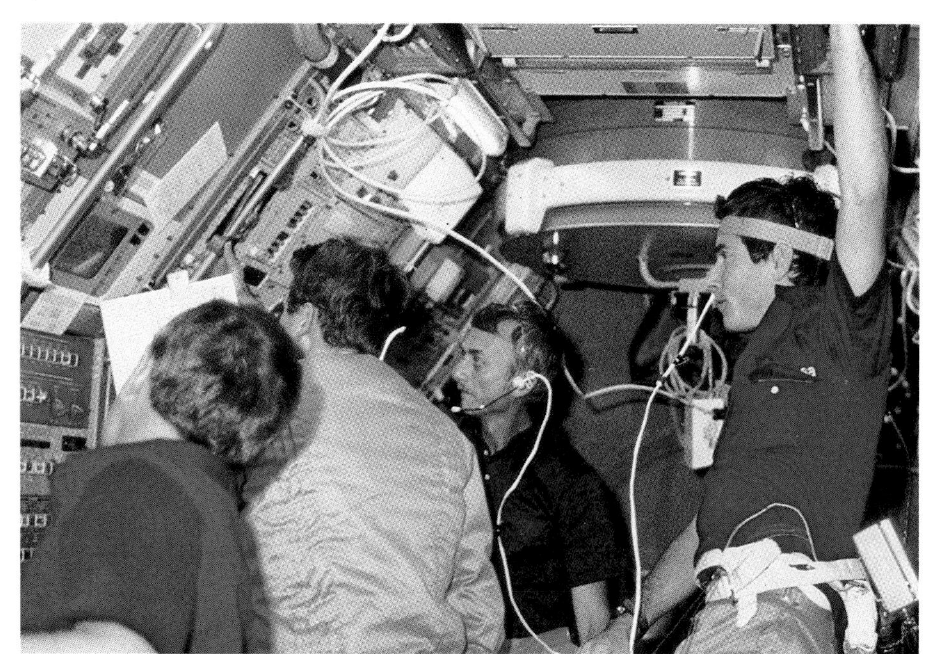

Sometimes Spacelab seemed rather crowded, as when the crew gathered around the only television monitor in the laboratory. Left to right are Robert Parker, Byron Lichtenberg, Owen Garriott and Ulf Merbold.

Spacelab and its connecting tunnel fill most of the payload bay of the Space Shuttle. The vertical line to the rear is the tail of the orbiter.

Ulf Merbold of the European Space Agency carries out a heating experiment on the Spacelab 1 mission.

By all accounts, Spacelab 1 was a success. So was the next flight of the Long Module on April 29, 1985, in spite of some unusual problems. For the astronauts on board, the flight provided some beautiful sights, at least for a while. For the first time, a Shuttle passed through an aurora, a glowing band of particles from the Sun electrically charged by the Earth's magnetic field.

In order to continue the space-sickness research of the first mission, small animals were carried in a specially built $10 million animal-holding facility. They included twenty-four laboratory rats and two squirrel monkeys. Each time the animals were examined for signs of space sickness, "clouds" of dry food particles were released into the cabin. Soon the crew had to wear surgical masks to prevent infection caused by breathing in the particles. Vacuum cleaners were used to clean up the mess, but some of the particles made their way up to the Shuttle flight deck. Ground controllers overheard mission commander Bob Overmyer asking in the background, "How many years did we tell them these cages would not work?"

LDEF

Looking something like a Chinese lantern, NASA's Long Duration Exposure Facility (LDEF) was offered to scientists as a new opportunity for long-term experiments in space. At the time, NASA had no idea just how long a time LDEF would really provide.

Previously, scientists who wanted to conduct experiments in space that would take months to complete had to be extremely clever in their approach. Only two avenues were open to them. One was to get the experiment accepted on board a manned space station like NASA's Skylab or the Soviets' Salyut stations. The other was to have it placed on board an orbiting satellite. In both cases, competition for limited orbital space was fierce among thousands of scientists, each of whom felt his or her experiment was the most important.

Then, a third possibility opened up, with many opportunities for experimentation. In 1984 NASA's Long Duration Exposure Facility offered scientists new opportunities for long-term experiments in space. LDEF is a twelve-sided structure 4.3 m (14 ft) in diameter and 9.1 m (30 ft) long. It is designed to fit neatly into the payload bay of the Space Shuttle.

The many sides of LDEF are divided into eighty-six trays, each about the size of a large and deep dresser drawer. Scientists are assigned one or more drawers for their experiments. It is up to them to design the experiments to fit the drawers precisely.

The Long Duration Exposure Facility was photographed just moments after the remote manipulator system arm of the Space Shuttle released it. The satellite holds more than fifty scientific experiments in small bays spaced around its exterior.

Two views of the Long Duration Exposure Facility as it was about to be lowered into its payload canister for transportation to the launch site, where it was installed in the Space Shuttle *Challenger.* Some of the scientific experiments are clearly visible in the LDEF's small bays.

LDEF is a relatively inexpensive way to conduct space experiments. LDEF has no experimental data communication system, and it doesn't need an attitude-control rocket system to keep it precisely aimed. At the end of an LDEF mission in space—normally lasting about a year—the Shuttle comes back and grabs it with its mechanical arm. It carefully places it in its payload bay for the return trip to Earth. Back on the ground, the experiment drawers are opened and presented to the scientists for study.

The first LDEF, still in orbit, is carrying fifty-three scientific and technological experiments that were provided by more than two hundred scientists from around the world. Some of the experiments are merely to capture cosmic dust, radiation and gas samples for analysis on Earth. Others are to determine the effects of exposure in space on various materials, coatings and electronic parts to find out how to build more durable space hardware in the future. Still others look at technologies for producing electric power in space and for controlling heat. One especially interesting experiment involves tomato seeds. Twelve million seeds were placed inside one of the trays. It was thought that cosmic radiation might change some of the seeds and produce interesting results when they are grown back on Earth. The seeds will be distributed to schools so that students will have an opportunity to experiment with them and understand the effects of space on them.

The first LDEF flight began on April 7, 1984, and was planned to remain in space for about eleven months. However, changes in Shuttle payload priorities delayed its March 1985 pickup. Before it could be retrieved, the Shuttle *Challenger*'s explosion in 1986 sidetracked all Shuttle launches until safety problems could be corrected. Instead of the eleven months participating scientists were promised, their LDEF experiments may remain in space six years before they are returned.

TRASH CANS

The payload bay of the Space Shuttle is huge. In its 318 cu m (10,600 cu ft) of space, there is room for several satellites and a variety of large pieces of scientific apparatus. Mission planners are careful to fill up the payload bay to make each mission as efficient as possible. However, no matter how carefully planners do their jobs, there is always some unused space left over.

Years before the first Space Shuttle actually flew, mission planners recognized the problem of wasted space and came up with a novel way to put it to use. They created a special class of space payloads called the Small Self-Contained Payloads Program. It was nicknamed Getaway Specials (GAS). Furthermore, they decided to offer these small payloads to anyone who wanted to fly a scientific experiment in space. Instead of costing millions of dollars to launch in space, a GAS payload would cost only thousands, even making it possible for individuals to send experiments into space.

In spite of their fancy name, Getaway Specials resemble nothing more than high-tech trash cans. GAS cans come in two sizes. The largest holds experiments with a mass of 91 kg (200 lb) and up to 0.14 cu m (5 cu ft) in size. The smaller GAS can holds payloads of 27.2 kg (60 lb) and 0.07 cu m (2.5 cu ft) in size.

An experimenter designs an experiment to fit inside the cylindrical GAS can. A rack is built to hold the equipment, such as a small crystal growing chamber, an electric power supply and a control device. It is then inserted into the can and sealed for space flight. NASA mounts the GAS can in a Shuttle orbiter payload bay along one of its walls or on a special partition called "the bridge."

(Left) A Getaway Special (GAS) canister being installed in the Shuttle payload bay. (Below) On some Shuttle missions a bridge is installed in the payload bay in order to hold over a dozen Getaway Specials at a time.

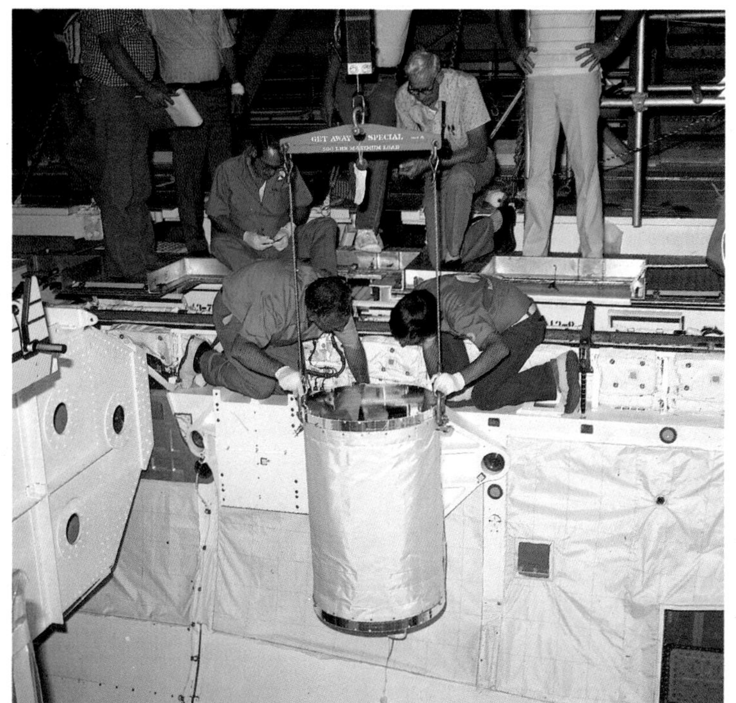

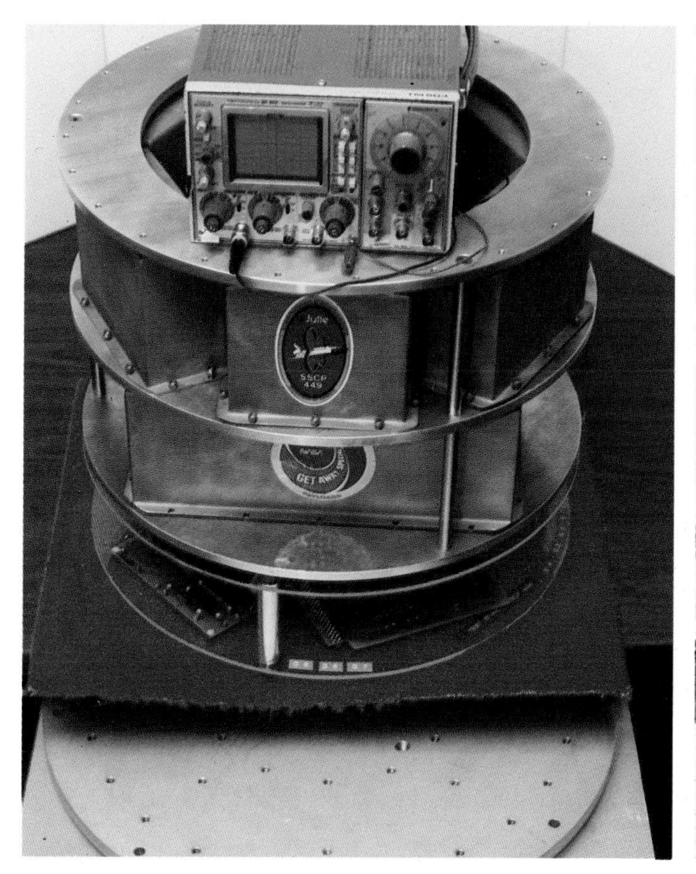

Once in space, the experiments inside the GAS cans are activated by an astronaut on the flight deck. With a small device about the size of a pocket calculator, the astronaut turns on the experiments. Control is limited to "on" and "off" and high and low speed or heat.

Otherwise, the experiments run automatically. Some GAS cans have lids that open in space, and they can even have a spring mount system that ejects a small satellite. Back on Earth, the GAS cans are unloaded and the experimenters take their experiments home for analysis.

Within limits, almost anything is possible with a GAS can. Experiments of all sorts can be carried. Many have been flown on Getaway Specials already, including those on crystal growing, snowflake formation, brine shrimp, lasers and medicine, blood typing and ants.

The limitations are that the experiments must be safe and must not pose a hazard to the orbiter, such as carrying something that might explode. The experiment must also be scientific or technological. In other words, someone cannot load up a GAS can with toys, stamps or pins and sell them back on Earth as space souvenirs.

There is also a practical limitation: A GAS can will be sealed up for days or weeks, so living things might not survive. Some researchers once suggested placing a small live pig in a GAS can for an experiment. It was not accepted for flight; imagine the problem of providing food, water and oxygen for a pig in a sealed GAS can for a week or more! Furthermore, who would keep the "pigpen" clean?

(Left) One of the payloads carried on the bridge on Shuttle flight 61C was JULIE. The name stands for Joint Utilization of Laser Integrated Experiments. JULIE, a collection of medical experiments, was sent into space by St. Mary's Hospital of Milwaukee, Wisconsin. (Above) Getaway Special canisters, large and small, are mounted together along the payload bay wall during the STS-7 Shuttle mission.

STUDENT SCIENTISTS

The third flight of the Space Shuttle *Columbia* in 1982 was beset with bugs. These weren't the everyday mechanical bugs that plague most complex equipment. These were real bugs: worker bees, velvet bean caterpillar moths and common houseflies. It was actually a space experiment with the name "Insects in Flight Motion Study." Remarkably, the experiment wasn't being sent up into space by a professional scientist. It was designed by Todd Nelson, a high school student from Minnesota.

Todd Nelson was one of the student winners of a contest that encouraged future scientists and engineers by allowing them to have their experiments flown in space. Nelson wasn't the first student to have such an opportunity. Many years earlier, a similar contest had been held for the Skylab space station launch. More than four thousand entries were considered before twenty-five student experiment proposals were selected. Ultimately, nineteen experiments actually flew on board the space station.

As a result of an experiment designed by Judith Miles of Massachusetts, Skylab astronauts had two pets on board, Anita and Arabella. Both pets were cross spiders, and Miles wanted to study spider web formation in space. The first webs spun in space were rather sloppy, but the spiders learned quickly and soon were spinning near-perfect webs.

In other Skylab experiments students studied bacteria growth, eye and hand coordination of astronauts, plants, liquids, astronomy and neutrons (atomic particles). Unfortunately, early troubles with the space station caused its interior to overheat, and some of the experiments were spoiled before repairs could be made. Nevertheless, other experiments were so successful that professional scientists constructed follow-up experiments for the U.S. and Soviet Apollo-Soyuz mission in 1975.

(Left) Todd Nelson makes a final adjustment in his container that held the first student experiment carried on the Space Shuttle.
(Below) In space, Nelson's bees, moths and flies reacted differently to the weightless environment.

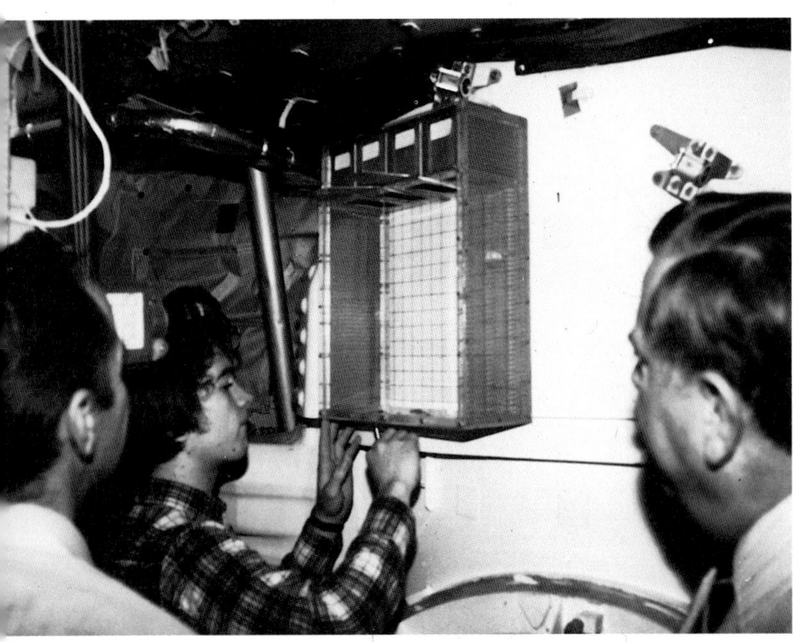

(Left) Arabella, a cross spider, learned how to spin a web in weightlessness in Judith Miles's Skylab experiment in 1973.
(Above) Michelle Issel holds her Shuttle student experiment; it investigated crystal formation in weightlessness.

Todd Nelson's Shuttle experiment initiated a long series of student experiments sent into space over the past few years. Experiments have covered a wide range of scientific fields. The space menagerie has been enlarged to include silkworms and earthworms. There have been more plant experiments and studies about liquids in weightlessness. The growth of crystals has also been a popular subject with students.

When the "bugs" of the *Columbia* were returned to Earth, Nelson examined television pictures of their flight characteristics. The weightlessness of Earth orbit had caused all sorts of problems for the bees. Some flapped their wings wildly, getting nowhere at all, while others just drifted around the box. The moths were better at adapting to space, but often they pitched their heads up and down as they traveled around the box. As might have been expected, the common housefly—a pest around the world—had little trouble flying in weightlessness. It was equally at home in space!

Dan Poskevich sent a colony of 3,000 bees into space to learn if the construction of honeycombs in space would be any different from the construction of a similar colony on the Earth.

MARTIAN LABORATORY

Exactly seven years after the first Apollo Moon walk, a glowing orange ball appeared high in the sky over the planet Mars. As the object slowed down in the thin air, a parachute billowed out and brought the object to a point 1,700 m (5,780 ft) above the desertlike surface. Suddenly, the parachute separated from the object and three rocket engines fired, bringing it down for a soft landing in the middle of what Earth astronomers call the Plains of Chryse. The surface mission of NASA's Viking spacecraft had begun.

Immediately, Viking radioed back to Earth that it was safely down and beginning its work. The message took nineteen minutes to travel across space. Soon, Viking's camera system sent back a picture of the rock-strewn Martian ground near one of its three landing feet. Later, pictures of the horizon showed essentially what a person would see if standing there. There were many rocks, ranging from dark gray to orangish gray in color, that were surrounded by fine soil. Some of the rocks had bubble holes and looked very much like rocks blasted out from volcanoes on Earth.

The Viking that landed on Mars in 1976 was the first of two Viking missions. The second landed on September 3 on the Martian Plains of Utopia. Their mission was to take photos of the Martian surface, to measure its atmosphere and weather, to detect "Marsquakes," to identify the chemicals found on its surface and, most important of all, to look for signs of life.

The Viking landers were not very large. Each was a hexagonal box about 0.5 m (1.5 ft) thick and about 1.5 m (5 ft) across supported on three legs. A dish antenna rose above the lander on a stalk and was aimed at the Earth. When needed, a metal arm—something like a steel tape measure—unrolled from a drum.

Artist's concept of the Viking 1 lander setting down on Mars. The artist made the drawing before the actual landing took place. Pictures sent back from Mars showed that the sky was pinkish and not blue, as here.

Once collected, soil and rock samples were transported by a tiny railroad system to three small laboratories inside the craft. The laboratories contained environment chambers, furnaces and analysis devices. One experiment tested the Martian soil to see if breathing things were present, and another looked for evidence of eating. Another experiment collected additional information about possible breathing.

While the life experiments were going on, other mini-laboratories were testing rock and soil samples to learn what chemicals they were made of. These and the life experiments would tell Earth scientists much about the Martian surface.

Early results of the life experiments indicated that living things were present on Mars, but later tests disputed those findings. In the end, scientists hoping to find Martian life were disappointed. It appeared that strange chemical reactions that only resembled life processes were taking place. Though the planet Mars may still contain life, the Viking science laboratories did not detect it.

(Left) The Viking 1 lander's soil-sampling scoop extends from the right, and a brush for cleaning the sampler jaws extends from the left in this picture sent back from Mars. (Above) Scientists were surprised at the large quantity of rocks and boulders found at the Viking 2 landing site. Early pictures sent back by the Viking 1 orbiter had indicated that this landing site would be much smoother.

Communication between the Viking landers and Earth is shown in this diagram. The lander can communicate by means of either a direct radio signal or a relay signal through the orbiter overhead.

FUTURE SCIENCE LABS

Ever since the beginning of scientific research, each answered question has opened a door behind which lie many new questions. Answering the new questions required designing new experiments and building new apparatus. Today, scientists are clamoring for more opportunities to conduct their experiments in outer space, on the Moon, on Mars, and anywhere else their imaginations take them.

Future science laboratories will be found under unusual conditions and locations. One will be located in permanent orbiting facilities in space. The Soviet Union already has such a facility. They call it Mir, from the word meaning "peace." Mir is a cylindrical space station with large solar panels that look like bird wings. It has docking ports, so that manned Soyuz spacecraft can dock at one end and unmanned Progress cargo spacecraft can dock at the other. Mir has become a permanent orbital laboratory for scientific and technological experiments.

The United States is planning its own permanent space laboratory for the early to mid-1990s. Its station will actually be an international effort. The European Space Agency is working on a module it calls *Columbus* to add to the station. Canada is working on a servicing center that will have robot arms for moving payloads around the outside of the station. Japan is also developing a module for the station. When it is assembled from pieces carried into space by the Space Shuttle, the station will serve both as a scientific research laboratory and as a space factory to make products that cannot be made as well on Earth. The factory will grow large, perfect crystals for use as semiconductors, produce new metal alloys, and purify biological materials for use as medicines and vaccines. A new product label, "Made In Space," will become common.

While NASA planners debate the final design of its new space station for the 1990s, scientists around the world are planning the experiments they would like to fly. This space station, an early design no longer being considered, featured many modules where scientists could conduct their experiments.

Future space stations with science laboratories will be located at great distances from Earth. This lunar facility will mine Moon rock for conversion into raw materials and oxygen for use in nearby laboratories and orbital space stations.

Earth orbit is the first step to more distant space laboratories. One obvious place for a new laboratory is the Moon. In the early twenty-first century, a permanent Moon colony may be established for both scientific research and materials processing. Apollo Moon landings found the Moon to be a good source of the metals aluminum and titanium. Mining the Moon makes good energy sense because the Moon's gravity is much lower than the Earth's. It will be much easier and cheaper to lift payloads off the Moon than off the Earth.

The Moon's far side will make an excellent location for radio astronomy equipment. Radio waves travel in straight lines in space, and the Moon's mass blocks them. So the only radio waves that reach the Moon's far side come from deep space.

Besides the Moon, there may be laboratories on the planet Mars, on asteroids, on the moons of Jupiter and even on other planets. The present and future science laboratories of space offer new opportunities for experimentation. They are limited only by the imaginations and abilities of scientists to ask questions.

One concept for an early Martian outpost serves as the base for long-term scientific studies leading up to more permanent laboratories and colonies.

IMPORTANT SPACE LAB DATES

October 4, 1957 Sputnik 1 is launched.

November 3, 1957 Laika, the first living creature in space, is launched into orbit.

January 31, 1958 Explorer 1 is launched.

November 29, 1961 Enos, the chimpanzee, is launched into orbit and later safely returns.

July 20, 1969 Apollo 11 lands on the Moon.

April 19, 1971 The Soviet Salyut 1 space station is launched.

May 14, 1973 The U.S. Skylab space station is launched.

July 20, 1976 The Viking 1 spacecraft lands on Mars.

April 12, 1981 The Space Shuttle is launched for the first time.

November 28, 1983 The first Spacelab flight.

April 7, 1984 The Long Duration Exposure Facility is released into orbit.

February 19, 1986 The Soviet Mir space station is launched.

GLOSSARY

Anita–A cross spider used in a Skylab student experiment.

Apollo–The project name of the three-man space capsule used for the U.S. Moon flights.

Apollo Lunar Science Experiments Package (ALSEP)–Automated science experiments placed on the Moon by U.S. astronauts.

Arabella–A cross spider used in a Skylab student experiment.

Columbus–A module being designed by the European Space Agency for the proposed U.S. space station.

Enos–Chimpanzee that orbited the Earth in a Mercury space capsule.

Experiment Module–A portion of Spacelab that can be joined to the Short Module to provide extra space.

Explorer 1–The first successful U.S. satellite.

Getaway Special (GAS)–A small self-contained payload that flies in a trash can-size container on board the Space Shuttle.

Habitation Module–The portion of Spacelab in which scientists can live.

Instrument Pointing System (IPS)–A system for the precise aiming of scientific instruments on the Spacelab pallet.

Laika–The Soviet dog that was the first animal placed into orbit above the Earth.

Long Duration Exposure Facility (LDEF)–A free-flying unmanned science laboratory that orbits the Earth for months at a time, then is retrieved by the Space Shuttle.

Long Module–The name given to Spacelab when the Short Module and the Experiment Module are joined together.

Lunar Orbiter–A satellite that orbited the Moon and took pictures of its surface to investigate possible landing sites for future manned missions.

Mir–The second-generation Soviet space station.

Pallet–An experiment rack placed outside Spacelab in the Space Shuttle's payload bay.

Ranger–A probe sent to the Moon that transmitted television pictures of its surface before crashing into it.

Salyut–The first-generation Soviet space station.

Short Module–The name of the Spacelab configuration when only one Habitation Module is carried on a mission.

Skylab–The only United States space station.

Small Self-Contained Payload–See Getaway Special.

Spacelab–A short-term space laboratory that was constructed by the European Space Agency for flights on board the U.S. Space Shuttle.

Space Shuttle–A winged spaceship that takes off like a rocket, orbits Earth like a spacecraft, and reenters the atmosphere and lands on Earth like an airplane.

Space Station–A permanent orbiting facility above the Earth that is occupied for varying amounts of time by crew members who carry out scientific and technological research.

Sputnik 1–The first satellite to successfully orbit the Earth.

Surveyor–A U.S. spacecraft that soft landed on the Moon.

V2–German rocket that carried bombs during World War II.

Van Allen Radiation Belts–Rings of electrically charged particles surrounding the Earth, trapped by the Earth's magnetic field.

INDEX